SpringerBriefs in Applied Sciences and Technology

SpringerBriefs present concise summaries of cutting-edge research and practical applications across a wide spectrum of fields. Featuring compact volumes of 50–125 pages, the series covers a range of content from professional to academic.

Typical publications can be:

- A timely report of state-of-the art methods
- An introduction to or a manual for the application of mathematical or computer techniques
- A bridge between new research results, as published in journal articles
- A snapshot of a hot or emerging topic
- An in-depth case study
- A presentation of core concepts that students must understand in order to make independent contributions

SpringerBriefs are characterized by fast, global electronic dissemination, standard publishing contracts, standardized manuscript preparation and formatting guidelines, and expedited production schedules.

On the one hand, **SpringerBriefs in Applied Sciences and Technology** are devoted to the publication of fundamentals and applications within the different classical engineering disciplines as well as in interdisciplinary fields that recently emerged between these areas. On the other hand, as the boundary separating fundamental research and applied technology is more and more dissolving, this series is particularly open to trans-disciplinary topics between fundamental science and engineering.

Indexed by EI-Compendex, SCOPUS and Springerlink.

More information about this series at http://www.springer.com/series/8884

José Afonso · Cristiana Bessa · Filipe Pinto ·
Diogo Ribeiro · Beatriz Moura · Tiago Rocha ·
Marcus Vinícius · Rui Canário-Lemos ·
Rafael Peixoto · Filipe Manuel Clemente

Asymmetry as a Foundational and Functional Requirement in Human Movement

From Daily Activities to Sports Performance

 Springer

José Afonso
Faculty of Sport
University of Porto
Porto, Portugal

Cristiana Bessa
Faculty of Sport
University of Porto
Porto, Portugal

Filipe Pinto
Fitness Hut Club
Rio Tinto, Portugal

Diogo Ribeiro
Fitrainer
Celorico de Basto, Portugal

Beatriz Moura
ME Development Clinic
Guimarães, Portugal

Tiago Rocha
Master Science Lab
São Félix da Marinha, Portugal

Marcus Vinícius
Master Science Lab
São Félix da Marinha, Portugal

Rui Canário-Lemos
University of Trás-os-Montes e Alto Douro
Vila Real, Portugal

Rafael Peixoto
University of Trás-os-Montes e Alto Douro
Vila Real, Portugal

Filipe Manuel Clemente
Instituto de Telecomunicações
Instituto Politécnico de Viana do Caste
Melgaco, Portugal

ISSN 2191-530X   ISSN 2191-5318   (electronic)
SpringerBriefs in Applied Sciences and Technology
ISBN 978-981-15-2548-3   ISBN 978-981-15-2549-0   (eBook)
https://doi.org/10.1007/978-981-15-2549-0

This Springer imprint is published by the registered company Springer Nature Singapore Pte Ltd.
The registered company address is: 152 Beach Road, #21-01/04 Gateway East, Singapore 189721, Singapore

Contents

Chapter 1
Introduction

The appeal of symmetry is deep. It conveys ideas of perfection, balance, and equilibrium. It makes reality seem rational and understandable. It soothes our spirits and provides a notion that the world is perfectly equitable and unbiased. And so, humans have searched for symmetry everywhere, and when nature did not abide, we imposed symmetric ideals. Mountains, rivers, and planets are irregular and asymmetric. But the human imagination has proven itself powerful and, through amazing conceptual jumps, attempted to impose symmetry onto nature. Triangles, circles, squares: all are completely imaginary figures, idealized concepts that have no connection to reality. Physics—as we will see—created several theoretical symmetries based on beautiful mathematical theories. But all these theories require the establishment of numerous symmetry-breaking events to be compatible with the known universe. So, while people search for symmetry, the key to existence is asymmetry.

Notwithstanding, tradition powerfully influences people's behaviors and conceptions, even if tradition is not always right. According to theoretical frameworks that address social conformism, a person will more easily accept a concept when it is practiced and stated by a group, even when the person is aware that this might not be the best action, thereby more easily discarding any responsibility or guilt toward the consequences [1, 2]. In this vein, common sense and daily practice in sports training and therapies tend to consider that symmetry is synonymous with health, and symmetry is highly recommended in the ninth edition of the American College of Sports Medicine's *Guidelines for Exercise Testing and Prescription* [3]. Still, in high-performance sports, asymmetries may be prevalent (e.g., dominant leg support or dominant leg strength) [4, 5]. Nevertheless, the presence of asymmetry is not always synonymous with increased injury risk for the athlete, as shall be explored in this manuscript.

In fact, the literature has shown that asymmetry-related pathologies occur only when certain thresholds are exceeded [6, 7], and there are also hints that symmetry may be detrimental to performance and health alike [5, 6]. Exercise and sports professionals may apply assessment tools to evaluate when asymmetries constitute

Asymmetry as a Foundational and Functional Requirement in Human Movement,
SpringerBriefs in Applied Sciences and Technology,
https://doi.org/10.1007/978-981-15-2549-0_1

a risk factor for injuries [8], but their interpretation requires caution [9–11], namely considering that the variability of functional anatomy is real in many aspects [12]. For that reason, the patterns and guidelines that are defined in those tests might not provide useful feedback in some cases.

The concept of asymmetry should not always be considered a negative issue, as its benefits depend on its magnitudes and the consequences regarding the functionality of the system. In fact, there are many examples in other scientific areas that emphasize the inevitability of asymmetries, namely in physics [13], chemistry [14], and biology [15–16]. So, despite symmetry being associated with a kind of idealized body perfection and a vital factor in some sports (e.g., bodybuilding), perhaps this concept should be reframed more broadly. Asymmetry should be understood in the context of an evolutionary process that starts in cosmology, expands into chemistry, is pervasive in biology, and, naturally, emerges in humans.

Our goal is to establish that asymmetry is required for all processes and is beneficial up to a point. Therefore, we shall present a rationale that includes the following sections: (i) from physics to chemistry: the role of symmetry-breaking; (ii) genetics and embryological development; (iii) structural and functional asymmetry in humans; (iv) asymmetry in daily activities; (v) asymmetry in athletic performance; (vi) injury prevention: from symmetry, to asymmetry, to critical thresholds.

Overall, we hope to promote a better, more refined understanding of asymmetry and to evaluate how valuable it is for existence, for life, for performance, and for sports. Of course, we are not so naïve as to believe that more asymmetry is always better. As with most phenomena in the real world, there are likely to be thresholds that, once crossed, make asymmetry detrimental. So, asymmetry is beneficial but must be limited to a certain magnitude. But what is that magnitude? Unfortunately, we are just starting to understand the thresholds, and they might be subject to a considerable amount of interindividual variability. Please, join us on this journey through a fascinating theme with many practical applications but also with very powerful philosophical implications.

References

1. Bond R, Smith PB (1996) Culture and conformity: a meta-analysis of studies using Asch's (1952b, 1956) line judgment task. Psychol Bull 119:111–137
2. Bond R (2005) Group size and conformity. Gr Process Intergr Relat 8:331–354
3. Pescatello L (2014) ACSM's guidelines for exercise testing and prescription, 9th ed. Wolters Kluwer, Lippincott, Williams & Wilkins
4. Bravi R, Cohen EJ, Martinelli A, Gottard A, Minciacchi D (2017) When non-dominant is better than dominant: kinesiotape modulates asymmetries in timed performance during a synchronization-continuation task. Front Integr Neurosci. https://doi.org/10.3389/fnint.2017.00021
5. Maloney SJ (2019) The relationship between asymmetry and athletic performance. J Strength Cond Res 33:2579–2593
6. Bishop C, Turner A, Read P (2018) Effects of inter-limb asymmetries on physical and sports performance: a systematic review. J Sports Sci 36:1135–1144

7. Jada A, Mackel CE, Hwang SW, Samdani AF, Stephen JH, Bennett JT, Baaj AA (2017) Evaluation and management of adolescent idiopathic scoliosis: a review. Neurosurg Focus 43:E2

8. Cook G, Burton L, Hoogenboom BJ, Voight M (2014) Functional movement screening: the use of fundamental movements as an assessment of function-part 1. Int J Sports Phys Ther 9:396–409

9. Maly T, Zahalka F, Mala L (2016) Unilateral and ipsilateral strength asymmetries in Elite Youth soccer players with respect to muscle group and limb dominance. Int J Morphol 34:1339–1344

10. Kollock RO, Lyons M, Sanders G, Hale D (2019) The effectiveness of the functional movement screen in determining injury risk in tactical occupations. Ind Health 57:406–418

11. Lisman P, Hildebrand E, Nadelen M, Leppert K (2019) Association of functional movement screen and Y-balance test scores with injury in high school athletes. J Strength Cond Res. https://doi.org/10.1519/JSC.0000000000003082

12. Zvijac JE, Toriscelli TA, Merrick S, Kiebzak GM (2013) Isokinetic concentric quadriceps and hamstring strength variables From the NFL scouting combine are not predictive of hamstring injury in first-year professional football players. Am J Sports Med 41:1511–1518

13. Thompson NW, Mockford BJ, Cran GW (2001) Absence of the palmaris longus muscle: a population study. Ulster Med J 70:22–24

14. Nagasawa M (2010) Cosmological symmetry breaking and generation of electromagnetic field. Symmetry Integr Geom Methods Appl 6:article 053

15. Strazewski P (2010) The relationship between difference and ratio and a proposal: equivalence of temperature and time, and the first spontaneous symmetry breaking. J Syst Chem 1:11

16. Stancher G, Sovrano VA, Vallortigara G (2018) Motor asymmetries in fishes, amphibians, and reptiles. Prog Brain Res 238:33–56

Chapter 2
Foundations of Asymmetry

2.1 From Physics to Chemistry: The Role of Symmetry-Breaking

Physics has established that reality relies on a series of phase transitions, resulting in symmetry-breaking events [1], and this can be extended to chemistry as well [2]. At a fundamental level, electrons and quarks, two elementary constituents of matter, have an intrinsic angular momentum—the spin—which, by convention, is said to be left-handed when rotation is clockwise and right-handed when it is anticlockwise. Only left-handed particles experience weak interaction or force, thus violating parity symmetry [3]. The electromagnetic field also emerges from a symmetry-breaking phenomenon, whereby it is no longer unified in an electroweak field [1]. In topological superconductors, initially symmetric crystalline structures produce spontaneous phase transitions that break time-reversal symmetry [4]. Asymmetry is also a fundamental geostatistical property, as it is a natural consequence of dynamic processes, such as land surface elevation and groundwater contamination [5].

Still, such asymmetries can be traced back to an even more fundamental level. Theoretically, the big bang should have produced equal quantities of matter and antimatter, but we can hardly find any antimatter in the observable universe, and this form of early, fundamental asymmetry between matter and antimatter is the foundation of our universe's existence [6, 7]. This foundational asymmetry, without which we would not be writing this manuscript, is hypothesized to have derived from quantum fluctuations that, in time, were exponentiated [8]. Furthermore, symmetry-breaking generates irreversibility, from which time itself emerges [2], and this can be observed at several levels (e.g., asymmetries in weak interactions, diffusion, quantum decoherence, and radioactive decay) [9].

Thus, physics and chemistry have shown that spontaneous symmetry-breaking is ubiquitous [2] and have established that the resulting asymmetries are more efficient from an energetic standpoint. Maintaining a symmetric state is energetically more demanding than maintaining an asymmetric state, and so most systems tend toward

Asymmetry as a Foundational and Functional Requirement in Human Movement,
SpringerBriefs in Applied Sciences and Technology,
https://doi.org/10.1007/978-981-15-2549-0_2

asymmetric states [10]. Attempting to balance a pencil on one's head is a perfect example of how symmetry is energetically demanding and highly unstable, resulting in a strong tendency toward more stable, asymmetric states. Therefore, it seems that reality itself would not exist if not for symmetry-breaking. Consequently, asymmetry emerges as a fundamental concept for existence and systemic organization and, thus, should be expected to translate to biological phenomena.

2.2   Genetics and Embryological Development

Genetics and the embryological development that follows are also grounded in symmetry-breaking. Symmetry-breaking is not limited to the chromosomal differences between the sexes of sexually reproducing species (an asymmetry that assures the survival of the species). According to Kawasumi et al. [11] and Shiratori and Hamada [12], the left–right asymmetry patterning in a mouse embryo requires four steps: (1) symmetry-breaking in the node around embryonic day 7.5, which occurs as a result of the leftward flow generated by the rotational movement of primary cilia in the node; (2) the transmission of an asymmetric signal (generated in the node) preferentially toward the left side of the lateral plate mesoderm (LPM); (3) the asymmetric expression of *Nodal* (a secretory protein that belongs to the transforming growth factor-beta or TGF-ß superfamily and plays a key role in signal transfer from the node to LPM and *Lefty2* in the left LPM; and (4) the situs-specific morphogenesis of the internal organs, which is mediated by the asymmetric expression of *Pitx2*.

More generally, genetically derived bilateral asymmetries are believed to be necessary for the developmental stability of living organisms [13]. Certainly, left–right asymmetries are fundamental in the formation of the body scheme across several phyla, occurring very early during embryologic development [14, 15]. The poor establishment of this expected asymmetry leads to left–right disorders, including situs inversus, heterotaxy, and dextrocardia, just to mention a few [16–18]. Of note, left–right asymmetry disorders of the heart represent a significant source of human heart disease [19].

Human left–right asymmetry disorders may be caused by genetic factors, environmental modifiers, and developmental stochasticity [20]. Relative to genetic factors, mechanisms like balanced translocations, duplications, inversions, insertions, small deletions, and complex chromosomal rearrangements are likely involved in left–right asymmetry disorders [20, 21]. As is clear, symmetry-breaking is decisive for proper embryological development and for human function. Humans, therefore, exhibit left–right asymmetry, most notably concerning the disposition and morphology of the internal organs [22]. In particular, left–right asymmetry is decisive in establishing the position of the thoracic and abdominal viscera, as well as the asymmetry inherent to each organ [23]. Starting from a laterally biased gene expression (i.e., certain genes express molecules that are preferentially rotated toward one side) [23], chiral molecules are produced and regulate the dynamics of the intracellular cytoskeleton, meaning that each cell already contains intrinsic molecular asymmetry [14, 15].

As an example of the complex interaction between genes and molecules in humans, the accumulation of serotonin (5-HT) on the left side of the embryo generates a unilateral *Nodal* expression, and other genes actively prevent *Nodal* expressions from spreading to the right side of the embryo [23, 24]. This correlates perfectly with the notion that asymmetry is physically and geometrically advantageous, as has been established in the previous section. Indeed, the distinction between left and right starts with a symmetry-breaking process [15]. Importantly, these embryological processes are common to all vertebrates and chordates [24], indicating that such asymmetries have a long history and have been preserved by natural selection.

References

1. Nagasawa M (2010) Cosmological symmetry breaking and generation of electromagnetic field. Symmetry Integr Geom Methods Appl 6:article 053
2. Strazewski P (2010) The relationship between difference and ratio and a proposal: Equivalence of temperature and time, and the first spontaneous symmetry breaking. J Syst Chem 1:11
3. Marciano WJ (2014) Quarks are not ambidextrous. Nature 506:43–44
4. Karnaukhov IN (2017) Spontaneous breaking of time-reversal symmetry in topological superconductors. Sci Rep 7:7008
5. Guthke P, Bárdossy A (2017) On the link between natural emergence and manifestation of a fundamental non-Gaussian geostatistical property: asymmetry. Spat Stat 20:1–29
6. Canetti L, Drewes M, Shaposhnikov M (2012) Matter and antimatter in the universe. New J Phys 14:095012
7. Perkins WA (2015) On the matter–antimatter asymmetry. Mod Phys Lett A 30:1550157
8. Kobakhidze A, Manning A (2015) Cosmological matter-antimatter asymmetry as a quantum fluctuation. Phys Rev D 91:123529
9. Ordonez G, Hatano N (2017) The arrow of time in open quantum systems and dynamical breaking of the resonance–anti-resonance symmetry. J Phys A: Math Theor 50:405304
10. Kibble TWB (2015) Spontaneous symmetry breaking in gauge theories. Philos Trans R Soc A Math Phys Eng Sci 373:20140033
11. Kawasumi A, Nakamura T, Iwai N, Yashiro K, Saijoh Y, Belo JA, Shiratori H, Hamada H (2011) Left–right asymmetry in the level of active Nodal protein produced in the node is translated into left–right asymmetry in the lateral plate of mouse embryos. Dev Biol 353:321–330
12. Shiratori H (2006) The left-right axis in the mouse: from origin to morphology. Development 133:2095–2104
13. Maloney SJ (2019) The Relationship Between Asymmetry and Athletic Performance. J Strength Cond Res 33:2579–2593
14. Brandler WM, Morris AP, Evans DM et al (2013) Common variants in left/right asymmetry genes and pathways are associated with relative hand skill. PLoS Genet 9:e1003751
15. McDowell G, Rajadurai S, Levin M (2016) From cytoskeletal dynamics to organ asymmetry: a nonlinear, regulative pathway underlies left–right patterning. Philos Trans R Soc B Biol Sci 371:20150409
16. Kosaki K, Casey B (1998) Genetics of human left–right axis malformations. Semin Cell Dev Biol 9:89–99
17. Levin M (2005) Left–right asymmetry in embryonic development: a comprehensive review. Mech Dev 122:3–25
18. Peeters H, Devriendt K (2006) Human laterality disorders. Eur J Med Genet 49:349–362
19. Kathiriya IS, Srivastava D (2000) Left-right asymmetry and cardiac looping: Implications for cardiac development and congenital heart disease. Am J Med Genet 97:271–279

20. Sutherland MJ, Ware SM (2009) Disorders of left-right asymmetry: heterotaxy and situs inversus. Am J Med Genet Part C Semin Med Genet 151C:307–317
21. Bisgrove BW, Morelli SH, Yost HJ (2003) Genetics of human laterality disorders: insights from vertebrate model systems. Annu Rev Genomics Hum Genet 4:1–32
22. Deng X, Zhou J, Li F-F, Yan P, Zhao E-Y, Hao L, Yu K-J, Liu S-L (2014) Characterization of nodal/TGF-lefty signaling pathway gene variants for possible roles in congenital heart diseases. PLoS ONE 9:e104535
23. Sadler TW (2019) Langman's medical embryology, 4th ed. Wolters Kluwer
24. Stancher G, Sovrano VA, Vallortigara G (2018) Motor asymmetries in fishes, amphibians, and reptiles. Prog Brain Res 238:33–56

Chapter 3
Structural and Functional Asymmetries in Humans

3.1 Gross Anatomical Asymmetries

There is evidence that skeletal asymmetries in the bones of the upper limbs have been intimately related to right-hand dominance since the dawn of the genus Homo; this type of asymmetry is also believed to have occurred in Neanderthals [1]. An anthropological study of 780 Holocene adult humans has shown that modern humans present bilateral asymmetry in the length and especially in the diaphyseal breath of long bones (e.g., the femur) [2]. Furthermore, the authors found a systematic right-bias in all dimensions for the upper-limb bones and a slight left-bias in diaphyseal breadth and femoral length. In another anthropological study of 509 Holocene adult humans, Auerbach and Raxter [3] found contralateral asymmetries in both the clavicle and the humerus, with greater asymmetries observed in diaphyseal breath than in length. Furthermore, the authors state that the asymmetries in diaphyseal breadth could be related to variations in the physical activities practiced by the groups.

In a study of 100 healthy humans (50 men, 50 women) without asymmetries in limb length, Shultz and Nguyen [4] evaluated bilateral differences. Bilateral asymmetries exceeded the measurement error for pelvic angle, tibial torsion, and navicular drop in more than 32% of cases. In the remaining measures (i.e., hip anteversion, standing and supine quadriceps angle, tibiofemoral angle, knee laxity, genu recurvatum, and femur length), left–right differences greater than can be accounted for by measurement error varied from 5 to 32% of the cases. So, even in subjects without bilateral differences in limb length, several anatomical asymmetries are still apparent. The authors thereby underlined the fact that measures taken from one limb may not reflect the reality of the contralateral limb.

The original photos presented below also illustrate different non-pathological bone conformations and engage in the theme of inter and intraindividual variations in the human anatomy that have movement-related consequences. In Fig. 3.1, two femurs are presented. They are extremely similar in several aspects, from their width to their rotation. The shafts of both femurs are also very similar. However, the orien-

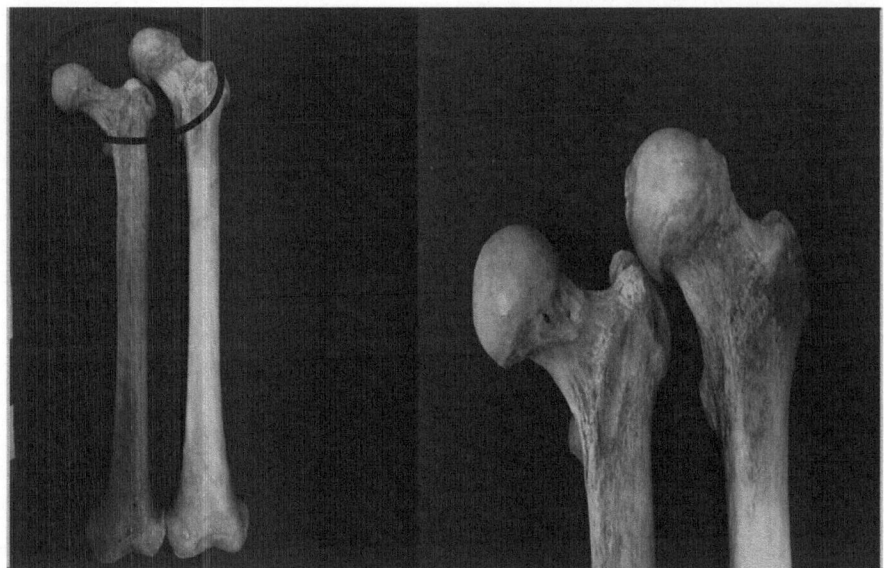

Fig. 3.1 Femur bones with similar distal and medium features, but highly distinct proximal features

tation and thickness of their anatomical necks differ. These differences may reflect different developmental paths in response to asymmetrical loading. True, interindividual variations at this level are likely superior to intraindividual variations, but these discrepancies should still be acknowledged.

In Fig. 3.2, we can observe two innominate bones with a similar basis but with very distinct rotations of the acetabulum and iliac wing. The first feature will impact, for example, the range of motion in terms of flexion and extension, with the left bone favoring flexion and the right favoring extension, and could also partially determine the degree of lumbar lordosis. The rotation of the iliac crest impacts the insertions of the gluteal muscles, among others, and impacts posture and gait.

In Fig. 3.3, two very similar scapulae present two very distinct spine orientations and acromion projections. The left-sided spine is oriented more superiorly, potentially promoting a greater degree of should abduction. Meanwhile, the right-sided spine is more likely to limit the range of motion, at least in terms of abduction.

Finally, Fig. 3.4 exhibits two almost equal humeri but with different heads. Differences in the head size and thickness may reflect distinct loading requirements throughout life and will also impact mobility.

Asymmetries can also be found in skeletal muscles. For example, a study of 300 Caucasian people (150 women, 150 men) between 18 and 40 years of age showed that 16% of the subjects exhibited the unilateral absence of the palmaris longus [5]. There are also registered cases of cadavers with double gemellus superior and double piriformis on one side of the body and a single gemellus superior and single piriformis on the other side; this has implications for the path of the sciatic nerve [6].

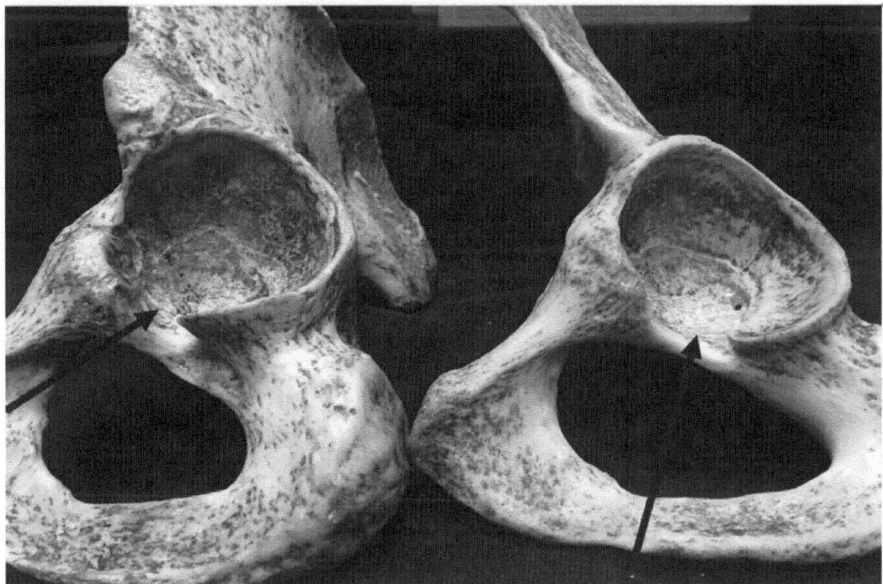

Fig. 3.2 Innominate bones with a similar basis, but with very distinct rotations of the acetabulum and also of the iliac wing

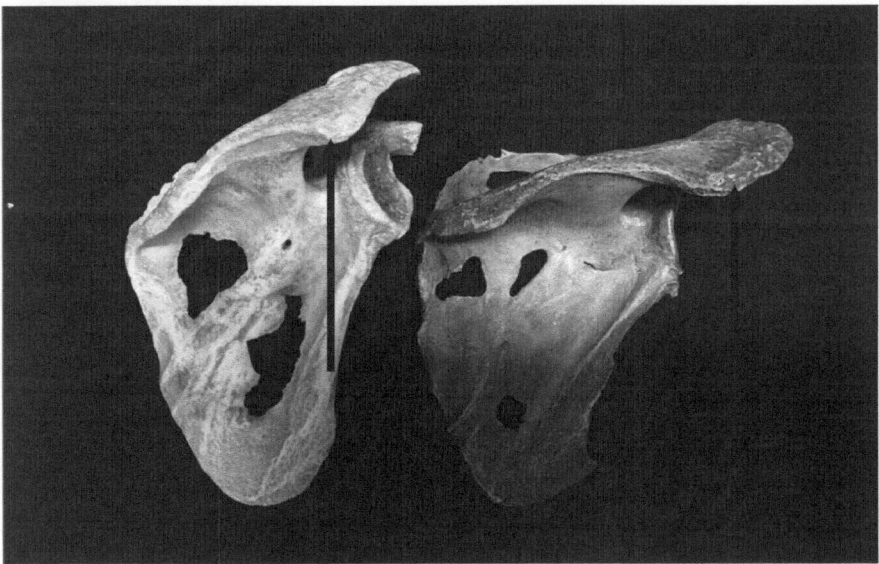

Fig. 3.3 Similar scapulae presenting two very distinct orientations of their spine and projections of the acromia

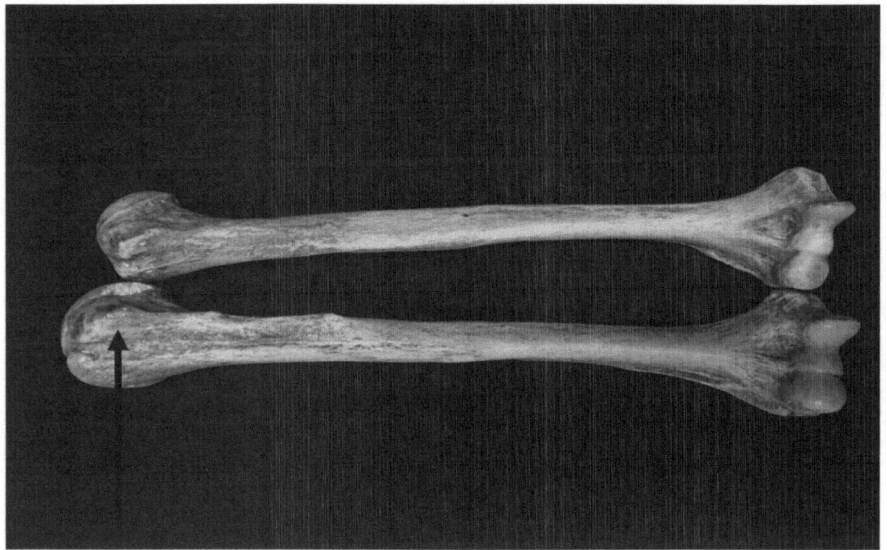

Fig. 3.4 Two humeri with distinct proximal features, but otherwise similar features

Incidentally, the primary author of this course has also found a case of unilaterally absent gemellus superior in one of its dissections (in the Faculty of Medicine of the University of Porto). In their thorough compilation, Tubbs et al. [7] present studies showing the unilateral absence of other muscles, such as the trapezius, serratus anterior, pyramidalis, and quadratus femoris. The partial or complete absence of a muscle is common (e.g., palmaris longus, serratus anterior, quadratus femoris) and isn't always associated with debilitating effects [8].

However, nowhere are gross anatomic asymmetries as expressive as in the internal thoracic and abdominal organs [9]. In fact, besides their asymmetric positioning, these organs are in themselves asymmetric, and when such asymmetry fails to develop, severe malformations occur [8]. The heart, for example, is biased toward the left side of the thorax (consequently also making the left and right lungs asymmetrical), and its chambers are highly asymmetric, reflecting the very distinct functional demands imposed on the left atrium and ventricle versus the right atrium and ventricle [10]. In fact, the circulatory system, as a whole, exhibits several major cases of bilateral asymmetry, as illustrated by the positioning and paths of the venae cavae and the aorta.

Another incredibly important structure is the diaphragm, which separates the thoracic and abdominal cavities; gives passage to the aorta, inferior vena cava, esophagus, and many other important structures; and is the main inspiratory muscle [8, 10]. Relevant to this work, the diaphragm has several asymmetric features: the three folia converging toward the central tendon are asymmetrical; the right side is usually oriented slightly upward due to the influence of the liver; the right crus terminates in L3 and is broader than the left crus, which goes only to L2; and even pathologies such as unilateral hernias are predominant on the left side (80%) [10]. Many more developmental asymmetries could be mentioned (the left thoracic duct and the position and structure of the stomach and the pancreas, etc.). Besides these, there are also unconventional asymmetries reported in the literature (e.g., the unilateral absence of the right or left common iliac artery, internal iliac artery, sigmoid sinus, denticulate ligaments, and olfactory bulb and tract) [7]. Together, these influence the paths taken by vascular, lymphatic, neural, and fascial tissues, implying asymmetric responses to movement. Of note, most of the asymmetries described in this paragraph were discovered only by chance – that is to say that the performance of human activities is not impaired, nor does pathology arise, from these asymmetries.

3.2 Neurological Asymmetries

Neurological asymmetries are a widespread feature of animals [11] and humans. Functional lateralization is not solely a byproduct of gross anatomical asymmetries, as they emerge very early during embryological development [12]. With regards to neurological features, perhaps the most well-known neuronal asymmetry is the brain's hemispheric dominance for language [11, 13], with the left brain being dominant in this regard in right-handed people. This notion has been reinforced by a recent meta-analysis [14]. Another meta-analysis revealed evidence for reduced language-related lateralization (i.e., smaller asymmetry) in patients with schizophrenia, especially in the temporal lobe [15], implying that asymmetry plays a positive, functional role in neuronal processing. More recently, Clark et al. [16] also reported decreased asymmetry in the temporal plane of adolescents with schizophrenia or schizoaffective disorder in comparison to a matched control group. These temporal plane asymmetries point to a wider role of structural and functional asymmetries in brain organization and are not limited to language [17]. For example, a meta-analysis has shown that resting asymmetries in the anterior prefrontal cortex appear to be important for properly regulating emotions and behavior [18]. A systematic review has demonstrated that more intense left prefrontal cortex activity is associated with positive psychological responses to exercise, including greater effect and energetic arousal and lower anxiety [19]. Overall, lateralized brain functions are thought to increase one's neural capacity [12].

Handedness is an asymmetrically distributed trait in humans [1] that can be partially explained by molecular mechanisms establishing right–left asymmetry early during embryological development [17], thus linking the chemical processes underlying cell development to neurological development. It is a clear-cut case of antisymmetry [9], even if hand preference and proficiency do not show a linear correspondence [1, 20]. Handedness is only one of a wide range of motor asymmetries in humans (as well as nonhuman species), and such asymmetries seem to play an evolutionary role [12]. Curiously, even therapists and physical trainers who endorse symmetry do not write their work with their nondominant hand. Here, society has for a long time actively promoted such bilateral asymmetries, with left-hand people being discriminated against and even called "sinister" in Latin languages. Even today, most human-built locations and objects are biased toward right-handed people. This is not the place to develop these arguments. They are mentioned here solely to illustrate that symmetry, while being widely sought out by humans, is actively discouraged in specific cases.

The neuronal regulation of visual search behavior is also prone to asymmetry. In one study, 92 right-handed participants were grouped into four classes: (i) strong right eye dominance; (ii) weak right eye dominance; (iii) strong left eye dominance; (iv) weak left eye dominance [21]. Eye dominance influenced saccade amplitude, with strong dominances producing more accurate saccades toward a target. In other words, greater asymmetries in eye function produced better results. Another study on 32 subjects with normal vision established that eye dominance occurs even under binocular conditions of visualization [22]. Eye dominance is an ancient and widespread feature in the animal kingdom [12]. Moreover, human eyes are controlled by six skeletal muscles (four rectus—superior, inferior, lateral, and medial—and two obliques—superior and inferior) and one smooth muscle (levator palpebrae) [7, 10]. Asymmetries in ocular control interfere with posture, possibly generating asymmetries concerning the positioning of the head and neck, although we should be cautious whenever attempting to establish cause-and-effect relationships.

It should be emphasized that neurological asymmetries are a feature of biological evolution, and hence, are not exclusive to humans. For example, pigeons' ascending visual pathways are top-down modulated asymmetrically [23]. Specifically, when presented with conflicting visual stimuli, the pigeons' right hemisphere is dominated by the left one, and the visual thalamus also exhibits profound asymmetries, as measured by extracellular single-cell activity using electrodes. These asymmetries might increase performance by reducing reaction time. Lateralization is also a feature of the mouse's auditory cortex, with the left auditory cortex specializing in the detection of specific vocalization features and the right auditory cortex being more generalist and better at integrative processing [24]. Also, Stancher, Sovrano, and Vallortigara [12] showed how strongly lateralized motor responses (i.e., marked left–right asymmetric preferences) in many (very) distinct species are positively associated with better reaction times (i.e., shorter latencies), thus enabling quick responses to environmental cues.

Interestingly, aerobic exercise has been shown to increase neuronal asymmetry. In a study of 12 healthy young adults, Hicks et al. [25] evaluated frontal alpha asymmetry, comparing the alpha power (8–13 Hz) of the right and left prefrontal cortices. Aerobic exercise effectively increased frontal alpha asymmetry until 38 min post-practice, with a relative increase in the activity of the left prefrontal cortex. The results also suggest that there is a minimum threshold of intensity that must be exceeded for changes in frontal alpha asymmetry to occur. Therefore, it seems that, above certain thresholds, exercise demands changes in neuronal activity, and, somehow, increased asymmetry in frontal alpha power is required.

References

1. Cashmore L, Uomini N, Chapelain A (2008) The evolution of handedness in humans and great apes: a review and current issues. J Anthropol Sci 86:7–35
2. Auerbach BM, Ruff CB (2006) Limb bone bilateral asymmetry: variability and commonality among modern humans. J Hum Evol 50:203–218
3. Auerbach BM, Raxter MH (2008) Patterns of clavicular bilateral asymmetry in relation to the humerus: variation among humans. J Hum Evol 54:663–674
4. Shultz SJ, Nguyen A-D (2007) Bilateral Asymmetries in Clinical Measures of Lower-Extremity Anatomic Characteristics. Clin J Sport Med 17:357–361
5. Thompson NW, Mockford BJ, Cran GW (2001) Absence of the palmaris longus muscle: a population study. Ulster Med J 70:22–24
6. Arifoglu Y, Sürücü HS, Sargon MF, Tanyeli E, Yazar F (1997) Double superior gemellus together with double piriformis and high division of the sciatic nerve. Surg Radiol Anat 19:407–408
7. Tubbs RS, Shoja MM, Loukas M (2016) Bergman's comprehensive encyclopedia of human anatomic variation. Wiley Blackwell, New Jersey, USA
8. Sadler TW (2019) Langman's medical embryology, 4th ed. Wolters Kluwer
9. Maloney SJ (2019) The relationship between asymmetry and athletic performance. J Strength Cond Res 33:2579–2593
10. Standring S (ed) (2015) Gray's anatomy, 41th ed. Elsevier
11. Corballis MC (2009) The evolution and genetics of cerebral asymmetry. Philos Trans R Soc B Biol Sci 364:867–879
12. Stancher G, Sovrano VA, Vallortigara G (2018) Motor asymmetries in fishes, amphibians, and reptiles. Prog Brain Res 238:33–56
13. Bradshaw JL (1988) The evolution of human lateral asymmetries: new evidence and second thoughts. J Hum Evol 17:615–637
14. Carey DP, Johnstone LT (2014) Quantifying cerebral asymmetries for language in dextrals and adextrals with random-effects meta analysis. Front Psychol. https://doi.org/10.3389/fpsyg.2014.01128
15. Sommer I, Aleman A, Ramsey N, Bouma A, Kahn R (2001) Handedness, language lateralisation and anatomical asymmetry in schizophrenia. Br J Psychiatry 178:344–351
16. Clark GM, Crow TJ, Barrick TR, Collinson SL, James AC, Roberts N, Mackay CE (2010) Asymmetry loss is local rather than global in adolescent onset schizophrenia. Schizophr Res 120:84–86
17. Brandler WM, Morris AP, Evans DM et al (2013) Common variants in left/right asymmetry genes and pathways are associated with relative hand skill. PLoS Genet 9:e1003751
18. Pence ME, Heisel AD, Reinhart A, Tian Y, Beatty MJ (2011) Resting prefrontal cortex asymmetry and communication apprehension, verbal aggression, and other social interaction constructs: a meta-analytic review. Commun Res Reports 28:287–295
19. Silveira R, Prado RCR, Brietzke C, Coelho-Júnior HJ, Santos TM, Pires FO, Asano RY (2019) Prefrontal cortex asymmetry and psychological responses to exercise: a systematic review. Physiol Behav 208:112580
20. Bishop DVM (1989) Does hand proficiency determine hand preference? Br J Psychol 80:191–199
21. Tagu J, Doré-Mazars K, Lemoine-Lardennois C, Vergilino-Perez D (2016) How eye dominance strength modulates the influence of a distractor on saccade accuracy. Investig Opthalmology Vis Sci 57:534
22. Johansson J, Seimyr GÖ, Pansell T (2015) Eye dominance in binocular viewing conditions. J Vis 15:21
23. Freund N, Valencia-Alfonso CE, Kirsch J, Brodmann K, Manns M, Güntürkün O (2016) Asymmetric top-down modulation of ascending visual pathways in pigeons. Neuropsychologia 83:37–47

24. Levy RB, Marquarding T, Reid AP, Pun CM, Renier N, Oviedo HV (2019) Circuit asymmetries underlie functional lateralization in the mouse auditory cortex. Nat Commun 10:2783
25. Hicks RA, Hall PA, Staines WR, McIlroy WE (2018) Frontal alpha asymmetry and aerobic exercise: are changes due to cardiovascular demand or bilateral rhythmic movement? Biol Psychol 132:9–16

Chapter 4
Asymmetries in Daily Activities

Many daily activities are typified by automated routines, most of which we are not even aware of. From brushing our teeth to holding and eating an apple, to opening doors, to driving, the number of hours spent daily performing highly asymmetric actions is considerable. Overall, human movement does not fit with the notion of perfect symmetry [1]. Writing is perhaps the best example of how asymmetries make our actions more efficient, as humans spend countless hours of their lives writing, especially during the school years. Writing consists of more than dexterity, as it also has implications for posture. For example, in a study conducted by Flatters et al. [2], the authors explored the relationship between body stability and manual dexterity in children ($n = 278$) aged 3–11 years. The results showed that postural control and manual control are interdependent and that the development of both postural control and manual control has a degree of task-specific codependency.

One activity that is performed every day is mastication, with thousands of chewing actions performed per day [3]. Analyzing the masseter muscles of 19 subjects under two different chewing conditions—soft (cake) versus hard (walnut)—Zamanlu et al. [4] showed that nearly 74% of the subjects preferably chewed with the right side when exposed to hard food. Even in the soft food condition, almost 60% of subjects chewed preferably with the right side. In another study, Rovira-Lastra et al. [5] selected 146 adults with natural and healthy dentition and found that 60% of the subjects were unilateral masticators. In horses, a lateral preference in mastication has also been reported, but here, the effects were highly individualized, without a clear trend for the species as a whole [6]. Instead of being restricted to mastication, motor lateralization is generally associated with feeding behavior in many different species [7] and, again, shows that this is not an exclusively human feature. It also illustrates that attempting to promote symmetry is, in fact, a deviation from the mechanisms that were actively "preferred" by natural selection.

Thus, laterality seems to be intrinsic to mastication regardless of whether the preference for the right side or left side is specific to the individual or to the species as a whole. It should be noted that mastication habits will likely asymmetrically

Asymmetry as a Foundational and Functional Requirement in Human Movement,
SpringerBriefs in Applied Sciences and Technology,
https://doi.org/10.1007/978-981-15-2549-0_4

develop the muscles of the face and neck (including the tongue), with potential implications for the thoracic region and, inevitably, to more distal regions as well. Certainly, mastication in adults with normal oral functioning and occlusion has been shown to asymmetrically activate the temporalis and masseter muscles of the left and right sides [8]. Therefore, and similarly to the case with writing, asymmetries in mastication potentially have broader implications for posture. Nevertheless, with reservations concerning statistical power, Hwang et al. [9] found that masticatory efficiency is associated with dynamical postural balance. Also, Izumi et al. [10] observed changes in masticatory patterns that depended on body posture (vertical versus reclined). Chewing patterns have also been associated with attention level and motor control [11], but the role of chewing asymmetry is unclear in this regard.

Gait constitutes another basic activity for most humans. In a study of 17 healthy young soccer players, Colombo et al. [12] verified that even though the players' gait produced highly symmetric vertical ground reaction forces, both the quadriceps muscle and the anterior cruciate ligament exerted right–left force asymmetries. That is, the resultant symmetry was produced by asymmetric mechanisms. Thus, even gross kinetic parameters can be deceiving, and subtler analyses may reveal hidden asymmetries. Using a combined gait asymmetry metric encompassing spatiotemporal, kinematic, and kinetic parameters, Ramakrishnan et al. [13] showed that overall asymmetry levels in gait vary due to several parameters that mutually cancel each other out, which reinforces the notion that even an approximately symmetric gait may be derived from locally asymmetric parameters. A certain extent of gait asymmetry is also normally present in schoolchildren [14], and most surprisingly, none of the asymmetry parameters were substantially affected by carrying a backpack or trolley. This suggests that the body habituates to a certain degree of asymmetry. Thus, when faced with additional challenges (e.g., carrying a backpack or trolley), the body alters its motor regulation to keep those asymmetries within the usual ranges.

Motor and postural asymmetries are indeed pervasive in the animal kingdom and not just in humans [7, 15]. A marsupial species called the red-necked wallaby (*Macropus rufogriseus*) presents a preference for using the left lower limb for tasks such as self-scratching, starting locomotion, and providing postural support [16]. A group of chimpanzees (*Pan troglodytes*) was shown to exhibit a preference for using the right hand for initiating walking, while the left hand was preferred for hanging and descending [17]. In response to a naturally asymmetric pronograde trunk orientation, small terrestrial birds called quails presented functional, left-skewed ground reaction forces when using bipedal locomotion, a strategy required for stable locomotion [18]. Limb preferences are also commonplace in tetrapods [7].

The ubiquitous presence of neuronal, motor, and postural asymmetries across several different species strongly suggests that asymmetry represents an evolutionary advantage. In addition, it has been suggested that such asymmetries might not change through training. For example, in a study of 60 polo horses, Pfau et al. [19] found that more than half of the horses exhibited asymmetries in their head movements, and the same rate was found for pelvic movements. Overall, almost 70% of the horses exhibited asymmetries beyond the levels accepted by the published guidelines. More importantly, none of the observed symmetry parameters changed after continued training. Still, reviews of the subject in the field of sports performance have provided contradictory accounts with regard to the changeability of asymmetry with specific exercise protocols (compare, for example, the systematic review of Bishop et al. [20] and the narrative review of Maloney [21]). Some asymmetry seems to be normal in human movement. Furthermore, most athletes possess a dominant limb for specific tasks, and these preferences may differ from task to task [1]. These issues will be further explored in the next chapters.

References

1. Ueberschär O, Fleckenstein D, Warschun F, Kränzler S, Walter N, Hoppe MW (2019) Measuring biomechanical loads and asymmetries in junior elite long-distance runners through triaxial inertial sensors. Sport Orthop Traumatol 35:296–308
2. Flatters I, Mushtaq F, Hill LJB, Holt RJ, Wilkie RM, Mon-Williams M (2014) The relationship between a child's postural stability and manual dexterity. Exp Brain Res 232:2907–2917
3. Verplancke K, De Waele W, De Bruyn H (2011) Dental implants, what should be known before starting an in vitro study. Sustain Constr Des 2:360–369
4. Zamanlu M, Khamnei S, SalariLak S, Oskoee SS, Shakouri SK, Houshyar Y, Salekzamani Y (2012) Chewing side preference in first and all mastication cycles for hard and soft morsels. Int J Clin Exp Med 5:326–331
5. Rovira-Lastra B, Flores-Orozco EI, Ayuso-Montero R, Peraire M, Martinez-Gomis J (2016) Peripheral, functional and postural asymmetries related to the preferred chewing side in adults with natural dentition. J Oral Rehabil 43:279–285
6. Parés-Casanova PM, Morros C (2014) Molar asymmetry shows a chewing-side preference in horses. J Zool Biosci Res 1:14–18
7. Stancher G, Sovrano VA, Vallortigara G (2018) Motor asymmetries in fishes, amphibians, and reptiles. Prog Brain Res 238:33–56
8. Kimoto K, Fushima K, Tamaki K, Toyoda M, Sato S, Uchimura N (2000) Asymmetry of masticatory muscle activity during the closing phase of mastication. CRANIO® 18:257–263
9. Hwang H-Y, Choi J-S, Kim H-E (2018) Masticatory efficiency contributing to the improved dynamic postural balance: a cross-sectional study. Gerodontology 35:254–259
10. Iizumi T, Magara J, Tsujimura T, Inoue M (2017) Effect of body posture on chewing behaviours in healthy volunteers. J Oral Rehabil 44:835–842
11. Hirano Y, Obata T, Takahashi H, Tachibana A, Kuroiwa D, Takahashi T, Ikehira H, Onozuka M (2013) Effects of chewing on cognitive processing speed. Brain Cogn 81:376–381
12. Colombo N, Vignaga F, Solari E, Merlo M, Manelli A, Negrini D, Moriondo A (2019) Gait screening of a population of young, healthy athletes by means of a portable, low-cost device unveils hidden left–right asymmetries in both quadriceps and anterior cruciate ligament forces. BMC Res Notes 12:366
13. Ramakrishnan T, Lahiff C-A, Reed KB (2018) Comparing gait with multiple physical asymmetries using consolidated metrics. Front Neurorobot. https://doi.org/10.3389/fnbot.2018.00002
14. Orantes-Gonzalez E, Heredia-Jimenez J (2019) Gait asymmetry and rating of perceived exertion: how are they influenced by carrying a backpack and pulling a trolley? Work 63:253–259
15. Corballis MC (2009) The evolution and genetics of cerebral asymmetry. Philos Trans R Soc B Biol Sci 364:867–879
16. Spiezio C, Regaiolli B, Vallortigara G (2016) Motor and postural asymmetries in marsupials: forelimb preferences in the red-necked wallaby (Macropus rufogriseus). Behav Processes 128:119–125

17. Morcillo A, Fernandez-Carriba S, Loeches A (2006) Asymmetries in postural control and locomotion in chimpanzees (Pan troglodytes). Am J Primatol 68:802–811
18. Andrada E, Rode C, Sutedja Y, Nyakatura JA, Blickhan R (2014) Trunk orientation causes asymmetries in leg function in small bird terrestrial locomotion. Proc R Soc B Biol Sci 281:20141405
19. Pfau T, Parkes RS, Burden ER, Bell N, Fairhurst H, Witte TH (2016) Movement asymmetry in working polo horses. Equine Vet J 48:517–522
20. Bishop C, Turner A, Read P (2018) Effects of inter-limb asymmetries on physical and sports performance: a systematic review. J Sports Sci 36:1135–1144
21. Maloney SJ (2019) The relationship between asymmetry and athletic performance. J Strength Cond Res 33:2579–2593

Chapter 5
Asymmetries in Athletic Performance

Based on the knowledge portrayed in the previous chapters, it would be surprising if asymmetries were not a factor in athletic performance. Ask a person to get ready to sprint, and that person will most likely always (and unconsciously) put the same foot forward. Ask a person to punch or kick a boxing bag as hard as possible, and their lateral preferences will tend to be very strong. These preferences are self-evident and can be easily accessed. But even in scenarios where it would be theoretically beneficial to perform actions with equal prowess on both sides, this does not occur in practice. In sports such as judo, boxing, or jiu-jitsu, most high-level competitors have well-established lateral preferences. Having two equally skilled sides would certainly prove to be a competitive advantage, and so we should ask ourselves why such symmetry is not common at the highest levels of practice, even when coaches stimulate bilateral practice in athletes starting from young ages. In elite-level basketball, players don't attempt three-point shots in equal percentages with both hands. Indeed, despite claims that bilateral asymmetries may be detrimental to sports performance, research does not entirely support this notion, as has been shown in the review of Maloney [1].

However, we contend that more subtle asymmetries may exist, even during apparently symmetrical exercise, something that has already been noted by Maloney [1]. As an example, consider the bench press and, for simplicity, imagine that the exercise is being performed on a guided machine that guarantees a bilaterally symmetric kinematical pattern. Despite an external appearance of symmetry, it is possible for one upper limb to contribute more force than the other. It is also possible for one limb to rely more on the pectoralis major, while the other limb is (relatively speaking) recruiting the anterior deltoid more intensely. The amount of force applied by each foot against the floor may also differ. Now, remember the previously mentioned asymmetries of the diaphragm, the effects of the liver, and so on—even if an exercise would appear symmetric, the internal effects on the thoracic and abdominal structures can never be symmetric.

Asymmetry as a Foundational and Functional Requirement in Human Movement,
SpringerBriefs in Applied Sciences and Technology,
https://doi.org/10.1007/978-981-15-2549-0_5

Lower limb asymmetries have indeed been reported in running, which is an apparently symmetric activity. In a study of 13 athletes with no previous injury, Girard et al. [2] created a context of five repeated 5-second sprints on a treadmill. Several bilateral asymmetries were observed: ~12.5% for propulsive power and horizontal forces; ~4% for resultant and vertical forces; ~6% for vertical stiffness; ~7.5% for leg stiffness; ~1.5% for swing time; and ~ 9% for aerial time. Interestingly, asymmetries in running may be beneficial during states of fatigue. Radzak et al. [3] analyzed kinematic and kinetic parameters in running in 20 healthy and physically active subjects. The subjects were evaluated in a rested state and after a two-stage treadmill fatiguing run. In both the rested and fatigued states, there were significant right–left differences in knee internal rotation, knee stiffness, loading rate, and adduction free moment. With fatigue, vertical stiffness, loading rate, and free moment became more symmetrical between the right and left lower limbs, but knee internal rotation and knee stiffness became more asymmetrical. In sum, fatigue induced greater symmetry in performance indicators, but others automatically compensated by increasing their asymmetries, again pointing to the greater mechanical efficiency of asymmetry over symmetry.

Anecdotally, the example of Usain Bolt's stride is also elucidative. As Longman reports in the *New York Times* [4], according to tests conducted by the Locomotor Lab of the Southern Methodist University, the sprinter's right leg is half an inch shorter than his left, and his right foot presents a 13% greater peak force upon ground contact than the left foot. However, the left foot remains 14% longer on the floor. These asymmetries seem to be working greatly for Usain Bolt and must be deemed functional to his anatomical and physiological characteristics. Notwithstanding, we also highlight that none of these asymmetries are over 15%, signifying that there are potential thresholds that, once surpassed, could have negative performance-related or pathological effects, a theme that will be discussed later section of this work.

The complexities of asymmetries in running have been superbly described by Ueberschär et al. [5], who analyzed three uniaxial components: axial, mediolateral, and anterior–posterior. Inertial measurement units were used to assess load parameters in 45 junior elite long-distance runners of the German national team (27 male and 18 female). Among other measurements, the authors have evaluated bilateral asymmetries in tibial and scapular accelerations. When considering average values, the authors reported a -2.5% difference in triaxial tibial acceleration, implying a minor dominance of the right leg. However, the standard deviation was 11.8%, meaning that each individual has relatively pronounced asymmetries while there is no clearly defined group tendency. For the scapulae, the asymmetry values were $+4.7 \pm 12.8\%$, again denoting the relevance of interindividual variability in asymmetry. The fact that the mean is positive, favoring the left scapulae, may reflect the utilization of the upper limb to counterbalance the right limb. When combining all values, asymmetry indexes are $9.0 \pm 8.1\%$, again denoting both the ubiquity of asymmetry and the considerable interindividual variability that exists. Analyzing the 95th percentile of the absolute asymmetry indexes, the authors concluded that tibial asymmetries below 13% in terms of 3D magnitude and below 22% in terms of axial acceleration could be considered normal in these runners even though they were all healthy. For the scapulae, that value was as high as 32%. These lateral asymmetries, therefore, might not be dysfunctional. Indeed, as the authors recognize, they may reflect an effective strategy for the runner to compensate for natural anatomical asymmetries, and therefore, attempts to correct running asymmetries could be harmful.

Asymmetry has also been identified in elite rowers [6]. The authors observed bilateral asymmetries within the range of 5–10% in the angle of the ankle joint and in the accelerations produced by the hip and knee joints. There were bilateral asymmetries of >10% in the resultant forces and in the acceleration of the ankle. These were associated with less pronounced inter-stroke variability (i.e., asymmetries promoted stable performance). Furthermore, kinetic asymmetries did not correlate with kinematic asymmetries or with asymmetries in the length of the lower limbs. Thus, asymmetry may manifest in very different ways, and morphological, kinetic, and kinematic asymmetries are not necessarily strongly associated.

In high-level gymnastics, asymmetries have also been reported. Analyzing beam routines of the 2014 B World Cup, Pajek et al. [7] observed that the gymnasts initiated ~43% of their takeoffs and landings with the right lower limb, while the left lower limb was used to initiate ~30% of the actions; the remaining takeoff and landing actions were performed with both legs simultaneously. Only four out of 19 gymnasts loaded the left lower limb more often than the right, denoting a clear group preference for loading the right lower limb and using it to take off or land. In a study of six high-level gymnasts, Exell et al. [8] performed kinetic and kinematic analyses of the upper limbs upon ground contact after a forward handspring on the floor. The six gymnasts presented significant kinetic asymmetries upon contact, and upper limb asymmetries could be traced back to the leading lower limb. Kinematically, the shoulders presented greater asymmetries than the elbow and wrist joints. As the authors highlight, these actions may appear symmetrical, but they are not.

In an analysis of a group of 29 elite female basketball players (under-16) who performed various tasks, including jumps, sprints, and changes of direction, Fort-Vanmeerhaeghe et al. [9] found significant differences in performance between the less-skilled and more-skilled limbs. However, this effect was highly individualized, and there was no group tendency. More interestingly, the participants' subjective identification of the dominant lower limb failed to coincide with the actual limb that performed better in nearly 50% of the cases. Because there are only two options (i.e., left or right), this means that the participants' awareness was not better than chance. Overall, this study highlights the need to use well-selected tests to evaluate dexterity, as this concept may be highly specific to the type of task being performed (also see Maloney [1]). This notion is further reinforced by the study of dos Santos et al. [10] of 20 multisport athletes (21 ± 1.9 years old), in which the asymmetries observed in the isometric midthigh pull were uncorrelated with the asymmetries observed in the 180° change-of-direction test. Furthermore, both asymmetries were uncorrelated with the actual performance observed during the tests. However, consistent with other studies, bilateral asymmetries were within reasonable ranges, with no athlete exhibiting differences of above 15%.

In volleyball, Lobietti et al. [11] have reported asymmetries in landing techniques. Having analyzed 12 matches from the Italian men's and women's first division, the authors compared landing actions with one foot versus those with two feet. Women landed more frequently with just one foot after a jump float serve, while men landed more often with one foot after spikes from positions 4 and 6. For men and women, quick attacks were related to landing more often on one foot. Overall, landing with one foot corresponded to 40% of the spike actions in men and 25.3% in women, 25.5% of serve actions in men and 20% in women, 48.8% of block actions in men and 41.3% in women, and 6.1% of setting actions in men and 17.2% in women. Therefore, even when considering supposedly symmetric actions, such as setting or blocking, landing occurs frequently with only one foot, and we should remember that athletes are exposed to such events many hours per week over the course of many years.

Analyzing 16 professional female soccer players (23.0 ± 3.8 years), Loturco et al. [12] examined the relationships between vertical asymmetries and performance in sprinting, change-of-direction, and muscle power tests. Asymmetries observed in the unilateral squat jump (SJ) and countermovement jump (CMJ) revealed differences between the dominant and nondominant legs, with the dominant leg exhibiting increased jump height, peak force, peak power, and landing force for both types of jumps. While bilateral jumping performance was strongly associated with the 30-m sprinting test and with the power exhibited in the jump squat, there were no significant associations between asymmetries in the unilateral jumps and any of the performance measures. This means that bilateral asymmetries in jumping ability did not impair speed- or power-related performance in these players.

In sum, asymmetries seem to be beneficial to sports performance, which is to be expected based on its physical and biological basis. This is the case even in apparently symmetrical actions. Notwithstanding, it is unlikely that asymmetry can increase indefinitely without producing negative effects. Therefore, enhancing sports performance may be followed by similar increases in the risk of injury. As such, we now turn our attention to the relationship between asymmetry and risk of injury.

References

1. Maloney SJ (2019) The relationship between asymmetry and athletic performance. J Strength Cond Res 33:2579–2593
2. Girard O, Brocherie F, Morin J-B, Millet GP (2017) Lower limb mechanical asymmetry during repeated treadmill sprints. Hum Mov Sci 52:203–214
3. Radzak KN, Putnam AM, Tamura K, Hetzler RK, Stickley CD (2017) Asymmetry between lower limbs during rested and fatigued state running gait in healthy individuals. Gait Posture 51:268–274
4. Longman J (2017) Something strange in Usain Bolt's stride. New York Times
5. Ueberschär O, Fleckenstein D, Warschun F, Kränzler S, Walter N, Hoppe MW (2019) Measuring biomechanical loads and asymmetries in junior elite long-distance runners through triaxial inertial sensors. Sport Orthop Traumatol 35:296–308
6. Fohanno V, Nordez A, Smith R, Colloud F (2015) Asymmetry in elite rowers: effect of ergometer design and stroke rate. Sport Biomech 14:310–322
7. Pajek MB, Hedbávný P, Kalichová M, Čuk I (2016) The asymmetry of lower limb load in balance beam routines. Sci Gymnast J 8:5–13

8. Exell TA, Robinson G, Irwin G (2016) Asymmetry analysis of the arm segments during forward handspring on floor. Eur J Sport Sci 16:545–552
9. Fort-Vanmeerhaeghe A, Montalvo AM, Sitjà-Rabert M, Kiefer AW, Myer GD (2015) Neuromuscular asymmetries in the lower limbs of elite female youth basketball players and the application of the skillful limb model of comparison. Phys Ther Sport 16:317–323
10. Dos'Santos T, Thomas C, Jones PA, Comfort P (2018) Asymmetries in isometric force-time charcteristics are not detrimental to change of direction speed. J Strength Cond Res 32:520–527
11. Lobietti R, Coleman S, Pizzichillo E, Merni F (2010) Landing techniques in volleyball. J Sports Sci 28:1469–1476
12. Loturco I, Pereira LA, Kobal R, Abad CCC, Rosseti M, Carpes FP, Bishop C (2019) Do asymmetry scores influence speed and power performance in elite female soccer players? Biol Sport 36:209–216

Chapter 6
Injury Prevention: From Symmetry to Asymmetry, to Critical Thresholds

Many postural and therapeutic protocols rely on reestablishing idealized levels of symmetry, specifically, left–right symmetry. For example, the ninth edition of the American College of Sports Medicine's *Guidelines for Exercise Testing and Prescription* states that a "training program should induce symmetrical and balanced muscular development" [1]. There is even a clinic called Symmetry Physical Therapy (https://www.symmetry-physicaltherapy.com/). Contrariwise, the rationale developed so far suggests that this perspective may be flawed. In this context, the systematic review of Bishop et al. [2] highlights the fact that we don't actually know the practical effects of attempting to reduce existing asymmetries. That is to say that even if such a reduction would be desirable, it may not be possible to achieve in the long run. In fact, motor asymmetries may be functional for performance and not necessarily associated with a greater injury risk [3]; remember that this feature has also been reported in polo horses [4]. Indeed, because symmetry is less efficient than asymmetry, most systems will spontaneously break symmetry [5], and human bodies that are forced toward more symmetric states might simply break symmetry at the earliest opportunity. In addition, most activities performed during sports performance will likely increase certain asymmetries [6].

Additionally, the systematic review of Bishop et al. [2] underlines that existing studies have relied on isolated assessments, and therefore, it is not known how functional asymmetries vary over time—for example, across a sports season. Coupled with the vastly discrepant methodologies used for evaluating symmetry [6], caution is advised when attempting to establish cause-and-effect relationships. An example illustrating the complexity of the relationships between asymmetries and injury comes from the study of Meyer et al. [7], in which 17 patients (post-reconstruction of the anterior cruciate ligament) were compared to 28 matched control participants. Although bilateral asymmetries were reported for landing mechanics from a drop vertical jump test and for knee joint laxity, the two were not related. Nine months after the surgery, the patients still presented mechanical strategies that unloaded the previously injured knee during bilateral landing, but this was unrelated to their increased degree of knee joint laxity.

Asymmetry as a Foundational and Functional Requirement in Human Movement,
SpringerBriefs in Applied Sciences and Technology,
https://doi.org/10.1007/978-981-15-2549-0_6

Another example comes from soccer, where Carvalho et al. [8] analyzed 159 professional players (75 from the First League and 84 from the Second League) with an isokinetic dynamometer for knee extension and flexion. There were no significant differences between leagues with respect to dynamic control ratios or bilateral asymmetry. There was also no relationship between competitive risk and injury risk. However, the authors concluded that coaches should promote work that minimizes bilateral asymmetries, even if their protocol does not address this issue. This conclusion is not supported by their own data. Furthermore, the authors referred to knee extension and flexion as if reflecting a relationship between quadriceps and hamstrings exclusively, thereby neglecting the roles of gastrocnemius, plantaris, sartorius, gluteus maximus and tensor fascia latae (via the iliotibial band), popliteus, and gracilis in knee extension and flexion. Regardless, a certain superiority of quadriceps strength over hamstrings strength in knee function (i.e., dorsal–ventral asymmetry) is accepted as normal.

In another study of 41 under-16 soccer players from a national team, Maly et al. [9] adopted a relatively moderate approach: when evaluating the players' knees with isokinetic dynamometers, the authors had already presupposed that asymmetries are a natural part of functional performance. Therefore, they considered that the athletes had bilateral asymmetries in strength only if the left–right difference was above 10%. The results revealed that bilateral asymmetries ranged from 19.5 to 31.7% for knee extension strength and between 36.6 and 51.2% for knee flexion. Over 73% of the players presented at least one of these asymmetries. Unlike in previous studies, the values of bilateral asymmetry presented are considerably elevated and again raise the question of how to define thresholds beyond which injury risk increases. Zvijac et al. [10] also failed to find isokinetic testing of the knee to be predictive of hamstring injury in football players.

Similarly, the systematic review and meta-analysis of Green et al. [11] established that isokinetic strength testing has limited predictive validity for hamstring injury risk. Having analyzed prospective studies, only 12 presented sufficiently good methodological quality to be included in the analysis. None of the 12 studies analyzed speeds superior to $300° \, s^{-1}$, which presents a challenge when translating such results to high-speed contexts. Globally, isokinetic testing had a very small effect on hamstring injuries. Combining this highly reduced efficacy with the considerable financial and temporal costs of performing these types of tests, the authors invite sports scientists to reconsider the utility of isokinetic testing for this purpose. Concerning knee flexion and extension, the existing consensus is that asymmetry exists in the sagittal plane, favoring extension over flexion. Still, these data demonstrate that athletes may require even greater imbalances, and those do not necessarily translate into increased risk of injury.

In cricket, the symmetry of abdominal muscle morphology has been associated with lower back pain. Gray et al. [12] analyzed 25 teenage cricket players specialized as fast bowlers, 16 of whom had low back pain, and nine of whom did not. The total combined thickness of external oblique, internal oblique, and transversus abdominis was superior on the nondominant side (i.e., at a diagonal with the dominant arm) in comparison with the dominant side. However, this was the case only for the subjects without lower back pain. Subjects with lower back pain exhibited symmetry in this respect. Specifically, the observed symmetry was explained by the reduced thickness of the nondominant side in comparison with subjects without lower back pain. This study provides another example of how the search for symmetry is flawed and might be associated with increased, not reduced, injury rates.

Wong and Cheung [13] analyzed the lower limb reaction forces of experienced rowers. No kinetic parameter differentiated subjects with pain ($n = 6$) from those without pain ($n = 19$). The asymmetries in the reaction forces of rowers with a history of back injuries were not superior to those observed in healthy rowers. Therefore, no relationship could be established between asymmetry and injury risk. Although the sample was small, the results support those of previously mentioned studies. In line with these results, Plastaras et al. [14] analyzed the hip abduction strength of 21 runners with early-stage patellofemoral pain and 36 matched, healthy controls. They verified that the bilateral asymmetries of runners with pain were not statistically different from those of the control participants.

In their prospective cohort study of 140 male US Air Force Special Forces, Eagle et al. [15] stated that bilateral strength asymmetries and/or unilateral imbalances were not predictors of ankle injury 365 days after testing when used as univariate predictors. When combined with body mass, though, both factors predicted ankle injury. There are, however, two major problems with the interpretations of the authors. First, the models were not predictive because they applied only to data that had already been collected. If the cut values can be produced to predict injuries in future samples, then they can be termed predictive. Unfortunately, this trend is often found in sports sciences, where the concept of prediction is applied but rarely with the proper design. To predict something, the model would have to predict the injury before it occurred. Here, no such thing happened. So, the authors should have spoken of association and not of prediction. Furthermore, when body mass alone was considered, the model showed very strong associations with injury. So, when asymmetries were combined with body mass, they also appeared to be associated with injury. In all likelihood, the very strong association (in this study) between body mass and injury was not impaired by strength asymmetries and imbalances. Ergo, the interpretation provided by the authors is not accurate.

So, it has now been well established that asymmetry is common and advantageous in several contexts, including sports performance. Moreover, it has been shown that cause-and-effect relationships between asymmetry and injury risk are difficult to establish. However, it should be underlined that the levels of bilateral asymmetry present in various research papers were confined to narrow ranges, usually $\leq 15\%$, though some highly superior values have been reported [9]. Accordingly, it is reasonable to assume a U-shaped curve, representing the following: (i) on the left side, a high injury risk is expected and is associated with symmetry; (ii) as asymmetry emerges, injury risk decreases; (iii) after reaching a critical threshold, increases in asymmetry would cause injury risk to increase once again.

For example, in their study of 150 military special tactics operators using self-reported knee injuries and testing for isokinetic knee strength, Eagle et al. [16] found that uninjured subjects had bilateral knee strength asymmetries lower than 20%.

Meanwhile, subjects with bilateral asymmetries above 20% had a 76–77% greater chance of having had a previous injury. Unfortunately, the systematic review of Bishop et al. [2] has shown that there are no consensual methods for evaluating the degree of inter-limb asymmetries and that the currently applied thresholds are arbitrary (i.e., they are not based on proper populational studies). The same arbitrariness has been denounced in the case of polo horses [4].

Furthermore, critical thresholds are likely distinct for athletic populations and general populations, as certain sports demand greater asymmetries than daily activities do [2, 6]. In other words, the same degree of asymmetry that is functional for the athlete might not be beneficial to the regular person. To illustrate our point, we analyze the work of Azevedo et al. [17], who observed 15 adolescent soccer players and 15 matched non-player controls. In a laboratory setting, performing a task as simple as standing barefoot on a pressure mat system, the soccer players presented plantar pressure asymmetries in the hallux, fifth metatarsal, and medial rearfoot, with greater pressure exerted by the non-preferred foot. These asymmetries were not verified in the control group. Despite the title of the article, which suggested that the risk of stress injuries in the foot of young soccer players was evaluated, there was in fact no evaluation of injuries or injury risk. In fact, none of the players had past injuries in the lower extremity.

This raises a very difficult problem: above a certain threshold, asymmetry is probably detrimental to a person's long-term health, though it might be beneficial for specific actions. Hence, complex ethical issues will be involved in any decision-making process, especially in the context of high-level performance. Another factor that renders such analyses difficult is the fact that asymmetry might be a cause in some cases and a consequence in others.

In the abovementioned paper by Eagle et al. [15], asymmetries were observed after the injury. Were they present before the injury? If so, were they causative, or were they unrelated to the injury? Or, are they merely a consequence? If so, is that a consequence of a positive protective mechanism, or is it an adaptation that may increase the risk of reinjury? These are very relevant questions to which we currently have no answer. Therefore, future research should address these complexities.

References

1. Pescatello L (2014) ACSM's guidelines for exercise testing and prescription, 9th ed. Wolters Kluwer, Lippincott, Williams & Wilkins
2. Bishop C, Turner A, Read P (2018) Effects of inter-limb asymmetries on physical and sports performance: a systematic review. J Sports Sci 36:1135–1144
3. Ueberschär O, Fleckenstein D, Warschun F, Kränzler S, Walter N, Hoppe MW (2019) Measuring biomechanical loads and asymmetries in junior elite long-distance runners through triaxial inertial sensors. Sport Orthop Traumatol 35:296–308
4. Pfau T, Parkes RS, Burden ER, Bell N, Fairhurst H, Witte TH (2016) Movement asymmetry in working polo horses. Equine Vet J 48:517–522
5. Kibble TWB (2015) Spontaneous symmetry breaking in gauge theories. Philos Trans R Soc A Math Phys Eng Sci 373:20140033

6. Maloney SJ (2019) The relationship between asymmetry and athletic performance. J Strength Cond Res 33:2579–2593
7. Meyer CAG, Gette P, Mouton C, Seil R, Theisen D (2018) Side-to-side asymmetries in landing mechanics from a drop vertical jump test are not related to asymmetries in knee joint laxity following anterior cruciate ligament reconstruction. Knee Surgery Sport Traumatol Arthrosc 26:381–390
8. Carvalho A, Brown S, Abade E (2016) Evaluating injury risk in first and second league professional Portuguese soccer: muscular strength and asymmetry. J Hum Kinet 51:19–26
9. Maly T, Zahalka F, Mala L (2016) Unilateral and ipsilateral strength asymmetries in Elite Youth soccer players with respect to muscle group and limb dominance. Int J Morphol 34:1339–1344
10. Zvijac JE, Toriscelli TA, Merrick S, Kiebzak GM (2013) Isokinetic concentric quadriceps and hamstring strength variables from the NFL scouting combine are not predictive of hamstring injury in first-year professional football players. Am J Sports Med 41:1511–1518
11. Green B, Bourne MN, Pizzari T (2018) Isokinetic strength assessment offers limited predictive validity for detecting risk of future hamstring strain in sport: a systematic review and meta-analysis. Br J Sports Med 52:329–336
12. Gray J, Aginsky KD, Derman W, Vaughan CL, Hodges PW (2016) Symmetry, not asymmetry, of abdominal muscle morphology is associated with low back pain in cricket fast bowlers. J Sci Med Sport 19:222–226
13. An WW, Wong V, Cheung RTH (2015) Lower limb reaction force asymmetry in rowers with and without a history of back injury. Sport Biomech 14:375–383
14. Plastaras C, McCormick Z, Nguyen C et al (2016) Is hip abduction strength asymmetry present in female runners in the early stages of patellofemoral pain syndrome? Am J Sports Med 44:105–112
15. Eagle SR, Kessels M, Johnson CD, Nijst B, Lovalekar M, Krajewski K, Flanagan SD, Nindl BC, Connaboy C (2019) Bilateral strength asymmetries and unilateral strength imbalance: predicting ankle injury when considered with higher body mass in US special forces. J Athl Train 54:497–504
16. Eagle SR, Keenan KA, Connaboy C, Wohleber M, Simonson A, Nindl BC (2019) Bilateral quadriceps strength asymmetry is associated with previous knee injury in military special tactics operators. J Strength Cond Res 33:89–94
17. Azevedo RR, da Rocha ES, Franco PS, Carpes FP (2017) Plantar pressure asymmetry and risk of stress injuries in the foot of young soccer players. Phys Ther Sport 24:39–43

Chapter 7
Concluding Remarks

The goal of this work was to convey that asymmetry, far from being a prejudicial concept, is a prerequisite for our existence, as it is found even in the form of matter–antimatter asymmetry [1, 2] and as far back as the creation of time itself [3]. Across a wide variety of levels, asymmetry emerges as an efficient solution [4] and provides the foundation upon which structures develop. Asymmetry is present in the rotation of subatomic particles [5], topological superconductors [6], and geostatistical processes [7]. Naturally, these asymmetries are necessary for proper gene expression and regulation, as well as for embryological development [8, 9]. Developmental, structural, and functional asymmetries are a feature of all vertebrates and chordates [10]. Therefore, reality fundamentally depends on asymmetry.

Consequently, asymmetry is ubiquitous in humans, both structurally and functionally. Examples include the asymmetries of bone dimensions [11], unilateral muscle absence [12], the positioning and structure of the thoracic and abdominal organs [9], and neuronal and motor lateralization [10]. Thus, humans perform many highly asymmetric daily routines (e.g., writing and driving). Even apparently symmetric actions such as mastication [13] and walking [14] are asymmetric, both kinematically and kinetically. Furthermore, asymmetries are the norms during athletic performance [15, 16], even for apparently symmetric activities such as running [17] and rowing [18]. It is only when asymmetry values surpass certain thresholds that pathology sets in [19]. What these thresholds are, however, remains largely an open question and might be highly variable, both from person to person and for the same person at different moments in time. It is clear, though, that symmetry is not inherent in humans. Moreover, when it does present itself, it might increase injury risk [20].

There is also a potentially vicious circle to be noted: structural asymmetries promote functional asymmetries; in turn, the repetition of asymmetric movements reinforces structural asymmetries. Indeed, Auerbach and Raxter [21] have suggested that dexterity may strongly influence anatomical asymmetries in the gross anatomy of the upper limbs. Existing asymmetries may be increased in sports, but current research lacks consistency and, therefore, a clear link between asymmetry and sports performance has not been established [16]. Overall, asymmetry is ubiquitous in humans

© The Author(s), under exclusive license to Springer Nature Singapore Pte Ltd. 2020   33
Asymmetry as a Foundational and Functional Requirement in Human Movement,
SpringerBriefs in Applied Sciences and Technology,
https://doi.org/10.1007/978-981-15-2549-0_7

and other animal species, but its degree-related effects should be the focus of future research [22].

So, what is the reason for this human obsession with symmetry? In a very interesting paper, Summerfeldt et al. [23] stated that the need for symmetry is an intrinsic component of the obsessive–compulsive experience. The authors enrolled 48 undergraduate psychology students based upon their scores on the Obsessive–Compulsive Core Dimensions Inventory (OC-CDQ), which also served as a way to divide them into two groups, each with 24 subjects. High-OC-CDQ participants reported feelings of incompleteness, engaged more often in symmetry-searching behaviors, and exhibited an increased preference for symmetry in images. In sum, the data suggest that searching for symmetry may reveal more about our feelings of incompleteness than they correspond to a desire for a symmetrical reality. Perhaps the search for symmetry can make certain processes more intelligible for us humans, but it could be a mere illusion: the perfectly symmetric ellipse is a circle and, despite being an idealized, platonic form, the ratio between its circumference and its diameter is π, an irrational number.

In any event, we should think of asymmetry as a continuum, ranging from zero (i.e., symmetry) to an undefined maximum value. In this respect, symmetry is but a very specific case in a range of possibilities (and a highly energy-demanding one, as we have seen), most of which involve some degree of asymmetry. The challenge is to understand when a given form of asymmetry stops being functional and becomes pathological. Such thresholds, if they exist, will likely present considerable interindividual variation, and possibly some intraindividual variation over time as well. Overall, asymmetry is fundamental to our existence and is a necessary feature of physics, chemistry, biology, and, of course, human movement and performance. Let us embrace asymmetry, but let us also keep its magnitude in check.

References

1. Canetti L, Drewes M, Shaposhnikov M (2012) Matter and antimatter in the universe. New J Phys 14:095012
2. Perkins WA (2015) On the matter–antimatter asymmetry. Mod Phys Lett A 30:1550157
3. Strazewski P (2010) The relationship between difference and ratio and a proposal: equivalence of temperature and time, and the first spontaneous symmetry breaking. J Syst Chem 1:11
4. Kibble TWB (2015) Spontaneous symmetry breaking in gauge theories. Philos Trans R Soc A Math Phys Eng Sci 373:20140033
5. Marciano WJ (2014) Quarks are not ambidextrous. Nature 506:43–44
6. Karnaukhov IN (2017) Spontaneous breaking of time-reversal symmetry in topological superconductors. Sci Rep 7:7008
7. Guthke P, Bárdossy A (2017) On the link between natural emergence and manifestation of a fundamental non-Gaussian geostatistical property: asymmetry. Spat Stat 20:1–29
8. McDowell G, Rajadurai S, Levin M (2016) From cytoskeletal dynamics to organ asymmetry: a nonlinear, regulative pathway underlies left–right patterning. Philos Trans R Soc B Biol Sci 371:20150409
9. Sadler TW (2019) Langman's medical embryology, 4th ed. Wolters Kluwer

10. Stancher G, Sovrano VA, Vallortigara G (2018) Motor asymmetries in fishes, amphibians, and reptiles. Prog Brain Res 238:33–56
11. Auerbach BM, Ruff CB (2006) Limb bone bilateral asymmetry: variability and commonality among modern humans. J Hum Evol 50:203–218
12. Thompson NW, Mockford BJ, Cran GW (2001) Absence of the palmaris longus muscle: a population study. Ulster Med J 70:22–24
13. Zamanlu M, Khamnei S, SalariLak S, Oskoee SS, Shakouri SK, Houshyar Y, Salekzamani Y (2012) Chewing side preference in first and all mastication cycles for hard and soft morsels. Int J Clin Exp Med 5:326–331
14. Colombo N, Vignaga F, Solari E, Merlo M, Manelli A, Negrini D, Moriondo A (2019) Gait screening of a population of young, healthy athletes by means of a portable, low-cost device unveils hidden left–right asymmetries in both quadriceps and anterior cruciate ligament forces. BMC Res Notes 12:366
15. Bravi R, Cohen EJ, Martinelli A, Gottard A, Minciacchi D (2017) When non-dominant is better than dominant: kinesiotape modulates asymmetries in timed performance during a synchronization-continuation task. Front Integr Neurosci. https://doi.org/10.3389/fnint.2017.00021
16. Maloney SJ (2019) The relationship between asymmetry and athletic performance. J Strength Cond Res 33:2579–2593
17. Girard O, Brocherie F, Morin J-B, Millet GP (2017) Lower limb mechanical asymmetry during repeated treadmill sprints. Hum Mov Sci 52:203–214
18. Fohanno V, Nordez A, Smith R, Colloud F (2015) Asymmetry in elite rowers: effect of ergometer design and stroke rate. Sport Biomech 14:310–322
19. Bishop C, Turner A, Read P (2018) Effects of inter-limb asymmetries on physical and sports performance: a systematic review. J Sports Sci 36:1135–1144
20. Gray J, Aginsky KD, Derman W, Vaughan CL, Hodges PW (2016) Symmetry, not asymmetry, of abdominal muscle morphology is associated with low back pain in cricket fast bowlers. J Sci Med Sport 19:222–226
21. Auerbach BM, Raxter MH (2008) Patterns of clavicular bilateral asymmetry in relation to the humerus: variation among humans. J Hum Evol 54:663–674
22. Corballis MC (2009) The evolution and genetics of cerebral asymmetry. Philos Trans R Soc B Biol Sci 364:867–879
23. Summerfeldt LJ, Gilbert SJ, Reynolds M (2015) Incompleteness, aesthetic sensitivity, and the obsessive-compulsive need for symmetry. J Behav Ther Exp Psychiatry 49:141–149